How To Build A Concrete Dome House

How to build the strongest, fireproof, tornado and earthquake resistant, concrete dome house.

By Jan Hornas

Copyright © 2000 by Jan Hornas

All rights reserved. No part of this book shall be reproduced or transmitted in any form or by any means, electronic, mechanical, magnetic, photographic including photocopying, recording or by any information storage and retrieval system, without prior written permission of the publisher. No patent liability is assumed with respect to the use of the information contained herein. Although every precaution has been taken in the preparation of this book, the publisher and author assume no responsibility for errors or omissions. Neither is any liability assumed for damages resulting from the use of the information contained herein.

ISBN 0-7414-0224-6

Published by:

Buy Books on the web.com
862 West Lancaster Avenue
Bryn Mawr, PA 19010
Info@buybooksontheweb.com
www.buybooksontheweb.com
Toll Free (877) BUYBOOK

Printed in the United States of America
Printed on Recycled Paper
Published January-2000

HOW TO BUILD A CONCRETE DOME HOUSE.

HOW TO BUILD THE STRONGEST, FIREPROOF, TORNADO AND EARTHQUAKE RESISTANT CONCRETE DOME HOUSE.

By
Jan Hornas

I would like to dedicate this book to everyone who wants to build a better house/home in which to live in.

Also I want to thank all those who designed and built those weak inefficient, drafty, obsolete houses I used to live in, which has spurred and inspired me into improving and creating houses that are more affordable, durable and energy efficient.

Jan Hornas   1999

How to Build a Concrete Dome House

by

Jan Hornas

TABLE OF CONTENTS

Why concrete	 4
Making plans	 13
Building the concrete dome	 14
Vertical wall	.... 21 - 22
Second floor, stairs	 23
Doors, windows	 23
Short history of ferrocement	 26
How to hire good workers and subcontractors	 27
Estimating cost	 28
Plans	.... 32 - 38

How to Build the Strongest, Fireproof, Earthquake Resistant Affordable Concrete Dome House.

Why Concrete?

Concrete is a perfect building material. It's fireproof, waterproof and maintenance free. Concrete houses will not burn, termites or any other rodents will not munch on it or even get in it. Concrete houses will last for a longer time than any other type of structure I can think of, and concrete will get even stronger with age as long as you mix and cure it properly. Concrete houses are bulletproof and with so many careless gun owners out there, accidents can happen. Concrete houses may be a lifesaver. I always say; Hope for the best, but be ready for the worst. In a concrete house you don't have to fear New Year's gunfire celebrations. Or any nearsighted trigger happy hunters who can't tell the difference between a house and a moose. Don't laugh, they are out there. Concrete houses can be built in almost any shape but the dome is most practical and strongest. The dome encloses the most space with the least material. It uses approximately 30% less materials than a square house of same size. It's much more energy efficient for even more savings. The concrete dome has no air leaks. The dome is much stronger than any other shaped house (except maybe a nuclear bomb shelter which has several feet thick walls). Not very affordable for most people. The dome is very strong because a round object such as an egg for example spreads the load, the weight placed on it evenly all along its sides without breaking easily. If you don't believe me take an egg test. Go ahead, go to the fridge , leave the beer in there or have one if it makes you feel stronger, grab an egg, hold it at the two longest points and try to crush it by squeezing it. Is it strong or not? Now if a thin little egg shell can be this strong, a concrete dome is certainly stronger than any other type of house, you will have to admit.

And when it comes to strong destructive winds such as a tornado, it will never crush the dome or knock it over, it will just slide over and around it. The round shape of a dome is so aerodynamic that wind has nothing to push against. A flat wall of a square house would have to be made from a very thick concrete to be equally strong.

Only windows could break of course, so it would be smart to make them from some unbreakable or bulletproof glass or plastic to feel completely safe.

A house is the biggest investment for most people. And the sad truth is, most if not all houses being built and sold out there are already obsolete. They are overpriced, they are weak, inefficient, junk. Maybe junk is a bit too strong a word you may say, but if a tornado force wind starts blowing strong at them, they will most likely crumble and fall down or up and blow away, that's junk in my opinion.

Also if you should accidentally burn your food on the stove, the cupboards, walls and ceiling will start burning in about two minutes flat and consequently burn the rest of the house down with it. That's obsolete. All stick houses, meaning anything built from 2 x 4 wooden studs and drywall, are outdated. Fortunately there is always better way of doing it.

My name is Jan Hornas. I am an inventor, artist, carpenter, and I have had my share of building all kinds of houses. Be it regular 2 x 4, I call them stick houses, because that is basically what they are made of, sticks, as compared to a log house for example. And if a tornado or hurricane starts blowing that's what will remain. A big pile of sticks. I always find it very interesting watching people board up their house's windows with plywood when the weather forecast calls for an approaching hurricane. Is that thin sheet of plywood going to make any difference, I don't think so. Why not build a stronger house in the first place?

I have seen or built many different kinds of houses and structures, mostly because I like variety, trying new things, trying to improve whatever I can, trying to make the world a better place, you could say, and hopefully do it a little better and if possible cheaper, as we all like to save a buck or two.

I have built some pretty unusual structures, anything from stack wall cordwood houses to straw bales covered with cement, also log and stone houses which all look and work quite satisfactorily up to a point that is ...

The problem is, logs will burn and are very heavy and difficult to move around and up and so are the stones.

And when it comes to 2 x 4 stick houses with their itchy insulation and paper thin walls ... forget it, I would not waste my money on that ...

My favorite building material is concrete, why? It can be built in almost any shape, will not burn, gets stronger with age, as long as it is mixed and cured properly and should last forever and a day, serving as a monument to your creativity. I told you I was an artist, so I guess houses are my expression.

The idea for a concrete house first came to me when I saw someone building a boat from concrete.

At first I couldn't believe it. Damn thing was only about an inch thick, the hull I mean. I figured if a 1 inch thick boat can withstand all that pressure of the ocean water and waves pounding over and over, than a house built the same way can certainly be strong enough to stand up to the worst weather, be it hail, heavy snow or a hurricane. And every time I have seen on the TV news that a tornado or hurricane destroyed some houses, I was wondering why don't they build stronger houses in those places. They must know that even if they rebuild those broken houses, the odds on being hit by a tornado next year again are just too damn high, so you will have to keep building and repairing those same houses forever and a day, maybe longer. Or until ... you smarten up and make it stronger. Like a concrete dome.

Even if you don't live in a Tornado alley, a concrete house dome makes more sense as you will soon see.

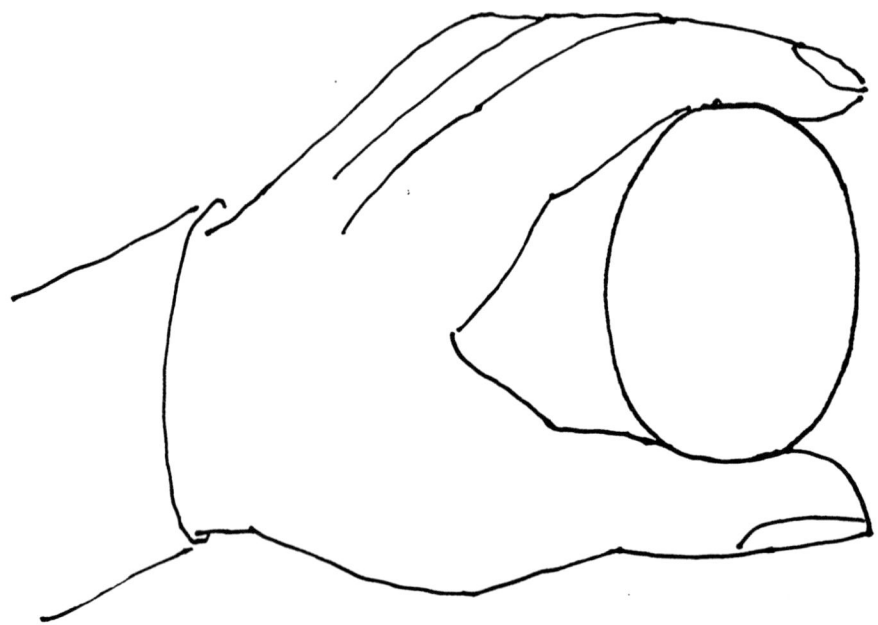

The proof is in the shape, nature does it perfect and with minimum materials. A round shaped dome is strongest. Learn from nature, the simplest most natural shape is best.

I am sure there will be some people who will worry that if you build a house that lasts forever and doesn't need any repairs all the workers will be unemployed and will starve. What a nonsense.

In this day and age when there is surplus of everything, especially food, when governments actually pay farmers to produce less food, and many people need to work only four days a week to live quite well, and I believe that in the very near future people will work even less as most of the work will be done by machines anyway, no one has to be poor or starving. Or live in a crackerbox-like firetrap that will need continuous fixing and repairs for as long as you own it. It is true though, if all houses were cheap and durable and did not need continuous maintenance, all the aluminum siding sales people would be out of work. So what, there is always something else to do. Besides do you really like that ugly siding? Do you like fixing it after some kid threw a baseball on it and broke it, or the wind blew it away?

Same goes for the rain pipes, they will rust and will need replacement, and cleaning all the time. Guess what, concrete dome home doesn't need any rain pipes. Tar paper shingles - that's another joke. Before you pay off your mortgage, you will need to replace them at least three times. More money down the tube. At least in some places such as Europe, Mexico and South America they are made of clay or concrete, which will last a lifetime, unless, here we go again, strong winds, earthquakes or just someone throwing a rock will break them; and when those shingles come sliding off the roof, you better stand very far away or wear a very big hard hat. Those shingles are hard and heavy. Better, but not the best idea, sorry.

In Europe they build houses from bricks, which is a lot better than the 2 x 4 stud system, they don't need studs, they don't even need any stupid, irritating fiberglass insulation, or plastic vapor barrier, or aluminum or plastic siding. The bricks they use have a whole bunch of vertical holes which act as an insulation, yes kids, dead air space is actually an insulator.

The walls are then plastered over with cement or stucco, it is simple, strong and effective. Those houses are so energy efficient that you don't need air conditioning, or a humidifier, or de-humidifier. And in the winter you can run the heater for a little while and then turn it off, because the house's walls will store and hold this heat for a very long time.

Unfortunately you cannot get those kinds of bricks in this country, don't ask me why. I don't know, but I have a pretty good idea why. We don't want a house that is low priced to build in this country. We have to buy and consume and waste more and more materials so the economy will keep going strong, what a bunch of crock. It sure is, we don't need to waste materials and in the process pollute the land, water and air.

In many European countries people already work 30 to 35 hours a week and retire at 50. Of course now you might say if people work less, they will buy less products, which is true, however, if people buy less then prices will come down eventually, so every thing will be pretty much the same, only better. How so?

For one thing we won't have as much ugly garbage to dispose of and you and I know that every big city on this planet has tons and mountains of garbage which they don't know where to put. Well anyway back to houses. Bricks are great, except for one big problem, in case of an earthquake or a tornado you better be somewhere else. When a brick or stone house falls down, it is an experience I would rather avoid.

Another way of building a house is from straw bales. Just mention this to some people, such as building inspectors and they will think you just escaped from the funny farm. Nothing funny about straw bales, though. You just stack them up, one over two and two over one, like bricks, spike them through with wooden sticks as you go up so they'll hold together, then cover them with wire mesh and plaster over with cement stucco. One great thing about bales, they are great insulators and very cheap building material. But when a strong wind starts blowing, I don't think it could really stand up against a tornado either.

Then there is the cordwood stack wall system of building. You just stack up short logs, 12 inches or so, lay them into cement and build the whole wall out of it. Then you plaster over inside and out with cement. It is very cheap and durable but in an earthquake it would probably crumble easily, just like any other weak house. But I guess, necessity being the mother of invention, people come up with all kinds of ideas for a shelter from the storm. In this country we have a lot of trees, so I guess that is why building houses from wood is so popular. I admit, well built log houses can look pretty nice but it is a really hard work, plus you have to treat the logs with some fire resistant varnish just to be sure it won't accidentally burn down.

Now we come back full circle to the concrete house. While a concrete house itself would be quite strong, if we make it round like a dome it would be so much stronger yet. So why, you may ask, did geodesic domes not become more popular? They are certainly an improvement over a regular square house, mostly in savings on quantity of building materials. It takes about 30% less material to enclose the same space with a dome than a square house.

One problem I can think of is that a geodesic dome is still basically a stick house made of 2 x 4s, these are arranged into triangles, hexagons, pentagons and what have you. All these different triangles of different sizes make it kind of complicated to build for an average woodworker.

Another big problem is sealing the roof as the whole dome is just one big roof. Tar shingles just were not designed for steep vertical surfaces. One easy way to solve the sealing problem is to make the whole dome out of concrete, thus having one continuous, uninterrupted solid roof over you. Yes, properly mixed cement will be waterproof when it hardens into concrete. So, if by building a dome home you could save around 30% on materials and possibly more, why aren't more people building concrete domes?

First, it is quite unusual to have a house of a different shape and looks than the rest of the population; many communities have strict restrictions on what shape you can build your house in their subdivisions, so you better be ready for a lot of attention and plenty of opposition from some people, especially building inspectors and contractors and trades people, should you ever need their services.

Now myself, I would not live in a square stick house that looks like all the other square boxes ... obsolete designs, requiring perpetual maintenance till the end of your days. I like to evolve slightly faster and build and live in a house that is an improvement over all existing designs. Life is too short to spend on just working and paying your mortgage and a whole lot of other little and big payments. I could not see myself doing that, especially if I had to live in a rat race subdivision, that is what it looks like to me, a bunch of rodents in a square maze. When people get stuffed into tight little boxes on a small piece of land , they turn into completely unpredictable mentally over-stressed "weirdos," to put it mildly.

I solved most of those problems by moving into the country, nice people, less restrictions on what you can build and clean air to breathe. So why aren't more people building concrete domes? They are so stuck in the rat race, so preoccupied to make money, to make ends meet and keep spending like crazy, they have no time to think that a house could be built in a better way. Also there just is not enough information anywhere on how to do it.

Until now, that is where this book comes in. I will try to make it short and simple so every average handy man and, yes, handy woman can easily comprehend what I am talking about. And if you get frustrated by uncooperative authorities, don't give up, even if you have to move to another location where people are more sensible, as you will see in the long run it will be worth every hassle, if you do everything right it will be a better house in every way.

A concrete dome can be built either way, on a concrete slab on grade, or over a basement. If you should build over a basement, I suggest that you place rigid insulation on the outside of the basement wall thus making it at least as energy efficient as the dome above it. It would not make much sense to build a nice airtight, energy efficient house-dome and have a cold, damp, uncomfortable basement under it.

Folks like me, who live in the country, have really no need for a basement, I'd rather build another dome or two as an addition as we have plenty of land to build on and no restrictions of any kind, plus I am figuring in the old age, or if someone disabled happens to come for a visit, you / they can get around the house easier if every thing is on just one level, instead of climbing up and down the stairs.

But if you must have a basement under your dome, yes, you can certainly build it. I would suggest making the floor also from reinforced concrete, if possible, that way you will further eliminate danger of fire.

I have seen too many fireplaces and wood stoves sitting directly, or very close, to wooden floors and or carpets, it was scary. It takes one little spark to fall down and ignite flammable materials like that very easily.

I like to have my fire well contained and isolated from every flammable material in the house, that way I can get a good night's sleep, even without fire alarm or smoke detectors. I know there is nothing to catch fire, so I don't even need insurance, more savings in my pocket.

Oh but you must have insurance because you have a mortgage? With a concrete dome you would and should pay much less since concrete does not burn. And hail will not damage any roof shingles, there aren't any. I don't' think even a flood would hurt the dome, except maybe the furniture.

Fire is I think the most common house destroyer, in a concrete house, I don't even have a smoke alarm. I believe my nose is a most reliable smoke detector anyway, if I smell even a whiff of smoke (sometimes even the best wood stove has a little back draft when the wind blows real hard) I am awake right away. Of course it helps to be a non smoker. It will keep your nose clean and your senses sharp. However, if you want to build a wooden floor, it can be done, too. You just have to make the basement wall wide enough to have a dome wall sitting on the outside edge of the basement wall and wooden floor behind it. You certainly cannot place concrete wall on top of the wooden floor joists, it would crush it. That is another plus for a concrete floor. It will cost a bit more but you can build the dome walls right on top of it. Plus, it will be a lot stronger.

After you have read all this book completely and understand all details about building this kind of dome house, I would suggest that it would be a good idea to start by building some small structure such as garage or shed or retaining wall, just to get used to working with cement and rebars. It is a demanding work even for a skilled, experienced builder. If you have no experience you will just need more practice, or hire someone who does.

Doors and windows. Somehow square windows just don't look right in a round house, I like to make windows round, or at least with an arch on top. Corners are the weak spots, where cracks may occur if the wall is not properly reinforced. However, if you want regular square windows, they can be used, too, without any problems. Make sure to plan well ahead and decide on size and shape of all windows and doors before you start building, once the cement hardens, the only way to add windows is with a concrete cutting saw. Not much fun, I should add.

Garage. It would be more practical to do in the dome shape with two ends flat-vertical, so you can install regular garage doors. Or one end can be attached right to the dome, depending which way it will fit on your lot.

There are few concrete dome manufacturers and they all seem to build their domes in very expensive, complicated ways, with need for inflatable forms, and sprayed insulation, and concrete sprayers which considerably adds to the price of building.

I would like to show you how to do it better and the cheapest way possible. Meaning low cost, not low quality.

My dome is built in a similar way as a ferrocement boat would be, the only difference is it has insulation, because every house needs it. If you build on some tropical island where the weather is comfortably warm and not too hot, you could probably do it without insulation.

Ferrocement means concrete reinforced with steel rebars, the thickness of the bars depends on the size of your structure. After you draw your plans (don't worry, it is easy) you should have an engineer to figure out what size of rebar to use. You will need an engineer's stamp of approval of the dome structure being strong enough to satisfy your local building code before they give you a building permit. If you don't need a building permit for whatever reason, I can only suggest to build a small section of the wall, maybe 4 x 4 feet long, let it cure properly, then place it horizontally, supported at ends only, and load it up with some heavy weights to see how much pressure it can take. Where I live, we need 60 lbs / square foot, a concrete dome can stand considerably more. You could and should also first look up some books on ferrocement boat building to get a general idea of the thickness of bars used. For example, if a 40-foot boat uses ½ inch rebar spaced 6 inches apart, it should be strong enough for a 40 foot dome. I said should be, not necessarily will be, get an engineer to figure this out for you.

So how do you build a ferrocement dome house? Very simply, start at the bottom and go up. I wish it was that easy. First you have to decide what exactly your needs are and what you can afford to build. Be realistic. What rooms, how many, how big. Draw your floor plan on square checkered paper, with each 1/4 inch square being one foot in real size, you will easily know how many feet in a certain room without even using a ruler. To get a proper idea how big a room should be, just look around you where you live, measure different rooms for size and adjust for size and shape where need be.

Important is, keep it simple, place all rooms according to your needs. If you don't have enough space for all different rooms you can plan on using one room for several different activities. Such as living room can also be dining room bathroom can also be laundry room, garage can also be workshop or storage, etc. If possible, place one bathroom close to the front door as that is usually the first place to use when coming home.

Bedrooms are last to use, so should be out of the way, preferably on 2nd floor, and have their windows facing the yard or garden and not the busy and noisy street, to make sure you get a good night's sleep.

Make the kitchen and living area windows overlooking the yard or driveway and street so you can keep an eye on your kids playing whether you work in the kitchen or rest.

Build only what you need, but plan for additions later, if you need to expand the house or build an addition, leave door openings which can be easily opened later.

One important thing, if you place the bathroom on the second floor, make the floor waterproof and put a drain in the middle. You will save yourself a lot of water related problems.

An entrance should have a foyer with two doors which acts as an airlock, so you don't get all that cold or dusty air blowing all through the house as soon as you open the door.

If you want to burn wood you will need a dry place to store it. It is best to keep it out of the house as there may be a lot of bugs in it, but still close to the fireplace for convenience. Garage is usually best. Or build an outside shed.

Make big windows on the south and west side of the house. Make small windows or none on the north. South west is warm and sunny, north is cold. Build the kitchen, bathroom and laundry close together so you don't have to run too many long water lines all over the house. If you want to save water, get a composting toilet that does not use any water.

Drawing Your Dome's Plans

Before you draw the house plan you must obviously have a piece of land on which to build it. Then you can place and orient the house in the proper direction. The plan should show the top, side and front views of the house. Draw your lot's size and shape on a piece of paper (square checkered paper is best as you can simply measure the distances by counting the squares). Mark north, south, west, east directions and your driveway. Then mark where you want your house to sit. Take into consideration all the practical realities of daily life. If you place your house too far from the road to have a quiet, secluded, relaxing and peaceful life, you may have to get a 4-wheel drive to get to it, if it snows heavily or when it rains and the road turns into a mud field.

On the other hand, if you are too close to a busy road or street, consider the possibility of speeding cars or an eighteen wheeler losing control and plowing into your living room or kitchen. While the concrete dome is quite strong and would probably hold up such an attack, I would rather try to avoid it, by building some strong fence or planting some big trees in front of it. Then draw your dome plan accordingly.

Use some of the plans in this book or make your own or modify them if you need to. Add or reduce doors and windows if you want. Sometimes all that is needed is just to turn these plans over or upside down to make them fit a different lot or face the proper direction.

The plan view from the top should be enlarged big enough to see all the details well as you will have to write and draw your plumbing and electrical lines and outlets. Of you can also have two plans, one for electrical and the other for plumbing or carpentry, such as interior partitions and cupboards, etc.

If you think you may be building additional domes attached to the first one sometime in the future, such as a workshop or another garage, leave door openings for it now.

Contact your building department (where you will get your building permit) and ask what else needs to be included on the plans. Then include it.

Usually what is required is just the top, side, front view and a cross section of the wall and roof, also an engineering stamp of approval of the designs strength. The plans for carpentry work, electric wiring, plumbing, etc. is mostly for you or the trades people you will need to hire if you can't do it all yourself.

However, the building inspector also likes to check all this to make sure everything is acceptable to the building codes. If you do work that you are not qualified for, you can still do it but have a certified trades person inspect it later, just to be sure it is done correctly.

Building The Concrete Dome

It is almost the same as building any other house. You will need very good, experienced concrete workers or plasterers and a carpenter, plumber and electrician. What you don't need are people to install insulation, siding or shingles.

First you make your house plan. After very careful measuring the location where your house will sit, make the foundation. Do not forget to dig a hole or a ditch for the plumbing, sewers and water lines at this time.

When making your footings, before cement hardens, make vertical holes about 6 inches deep, spaced evenly apart, this can be anywhere form 6 to 12 inches, depending on the size of your rebars as required by the engineering calculations.

You did get an engineer to approve the design, right? If not you will be just guessing how strong it will be, so add more and bigger rebars and space them closer together. It is not really a nuclear science, just use common sense when in doubt make it stronger.

For foundations in a hurricane alley, you may want to increase the holding strength of rebars by inserting short two foot long bars with the lower end bent 90°, like a hook into the wet cement, let them stick out about a foot. When the cement hardens, you can tie the dome's bars to them. You don't need too many of these, place them approximately every five or ten feet. I don't think any tornado would push the dome off the foundation but if you are worried it might, this will make it just that much stronger.

When it comes to insulation, I figure two inches are plenty. If you look at your fridge, you will notice how little insulation it has and it still keeps the ice from melting even in a warm room. So if you use high quality insulation, these go by many different names, but even the low priced styrofoam will do the job. I like it because it bends easier, however, it is a bit softer, so you will have to be more careful not to break it. Whatever insulation you use, it is easier to use two layers and overlap these wherever there is a gap; the thick hard insulation is difficult to bend around some sharp curves so you will be better of using two or even three 1-inch boards than one thick board.

You could also inquire about a price for sprayed insulation if you can find someone who does it. If ind the price too high. If there are only one or two people within a thousand mile area, they tend to charge a bit too much because there is no competition. They figure if you need it that bad you will pay whatever they ask. They figured wrong.

Depending on the thickness of your insulation, measure from the outside edge of the footing, add one inch for exterior cement coating, then insulation, then make holes. Into these you will later put the dome's rebars.

So if you have 4-inch insulation plus 1-inch concrete outside, make those holes about 5½ inches form the edge, that is where your interior wall will be.

When you do the foundation, the easiest way to make a circle is to drive a stick in the middle, stretch a rope or tape from it then just walk around and draw a circle. Depending on the size of your dome, you will need either a stepladder or scaffolding, probably both.

Now you can start building the dome by putting together all the bars into what will be basically a cage of vertical and horizontal rebars. Depending on the height of the dome, you may have to tie or weld two or three bars end to end as they are only 20 feet long. Insert one end into the hole in the foundation, affix the other end temporarily on top of the scaffolding which should be right in the middle of the dome.

Run another bar from the opposite side and tie them together with wire. Continue with the rest of the bars, making sure they curve at a proper circle so it will look like a dome. A perfect circle is strongest and nicest I think, though en egg shape or ellipsoid would still be strong enough, too.

Use a long stick or 2 x 4 to push up on the bars and support it while another person ties it securely at the top. When you add more bars, it will get a bit crowded on top, they will be closer together so you will have to cut some of them shorter. If there is no place to tie the bars on top of the scaffold, place one horizontal bar and tie it curving in a circular shape and tie the rest of the bars to this.

Once all vertical bars are up tie on all the horizontal ones starting at the bottom and go up.

Don't forget to leave all openings for doors and windows at this time. Make sure to tie all the bars real good at each crossing. When that is done, you can tie in all electrical boxes, outlets (use the shallow ones) and either metal or plastic conduit. Tie all these to the bars.

You can run your wires now or later, it doesn't really matter. I would do it later after you have a roof over your head. Make sure not to forget any wires or plugs. If you do, you will have to run them on the interior wall. It would be OK in the garage or workshop but not very nice in a house.

After securing all electrical conduits in place, it is time to tie the wire mesh known in trade as a chicken wire as it is also used for making chicken cages and fences. Us the one with the smallest one inch holes or the cement will have nothing to hold onto. Use at least four layers of the wire mesh and make sure to stretch it real tight. Put two layers inside and two outside. If you can find some wire mesh that is stronger and has smaller holes, it would make the plastering job much easier. It would most likely cost a

When putting the rebars up, use the following order;

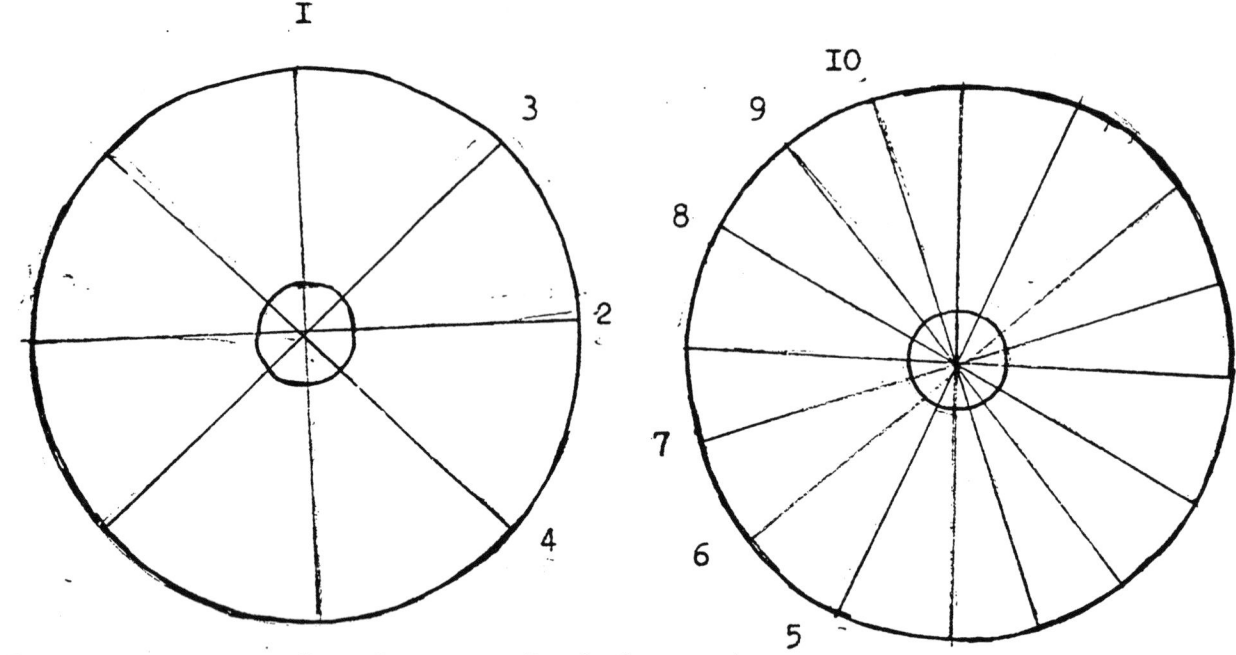

Leave openings for doors and windows at this point.
Once the bars start getting too close together up on top, you'll need to cut some shorter.

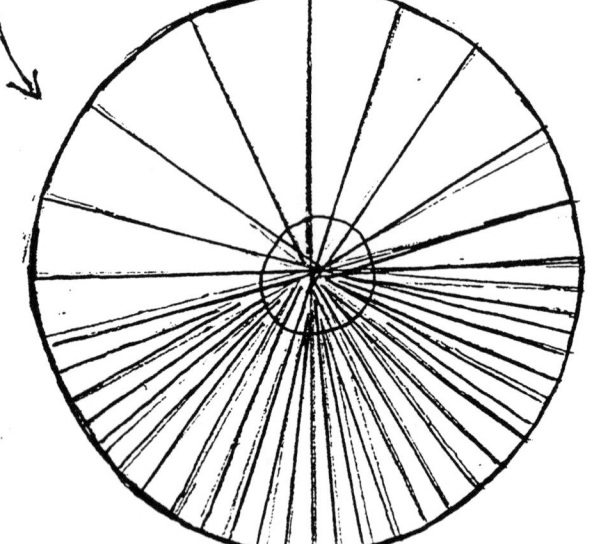

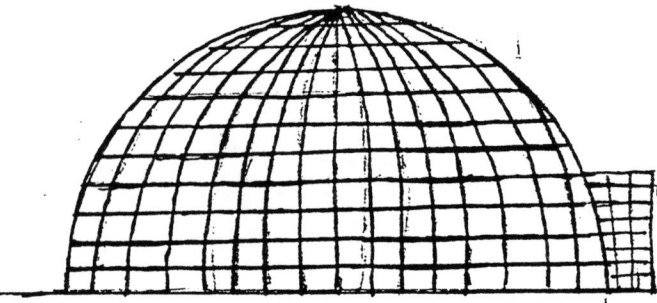

When all verticals are up, do the horizontal ones.

-16-

bit more though. The wire mesh has to be tied to all the bars every one or two feet to make sure it will all hold together. It is an awful slow tedious process but it needs to be done. Have a good supply of protective leather gloves, you will wear out quite a few.

Don't forget to cut out the mesh where the electric outlets are. Stick a piece of wood or insulation or something in the boxes so they don't get plugged with cement when you plaster the wall. Before you start plastering, make sure that you did not forget any openings for wires, plumbing, gas lines, etc. Check again to be sure.

Now when all the rebars and wire mesh are properly tied together and all wiring conduits are in the walls, you are ready for cement. Proper mix ratio is: one part Portland cement to three parts sand, that way it will be strongest. Do not use any gravel as it would be impossible to push into the mesh.

First a word of warning though, **do not get wet cement on your hands or anywhere else, it will eat your skin and flesh as it is extremely corrosive, just like acid.**

Be very careful not to splash any in your eyes, if that happens wash with clean water right away and get medical attention immediately. Wet cement is dangerous stuff.

Cementing the dome is very slow until you get a hang of it. Start from the bottom and continue all the way around with a "one trowel scoop high" layer, maybe six inches or so, have one person, the more skilled one, doing the inside and one the outside and one just mixing and carrying cement.

Do not try to go up too fast, it is best to continue in a horizontal direction all the way around just a few inches at a time. If you can't get a perfectly smooth finish with the first layer (and I don't think anyone can), don't get frustrated, just do the best and keep going on.

By the time you will get around for a second time, the first layer of cement should be stiff enough for you to finish it. It will take two or maybe even three passes before you get a nice smooth finish, so don't worry about the first layer being too rough, it is OK, as it will make the second coating stick better. Just make sure to wet it with water if it is dry already so that the second layer adheres well.

It is important to make a nice smooth finish only inside as the outside will be covered with insulation so don't kill yourself trying to make a perfect finish there. It is also important to make sure to cover all metal, rebars and wire mesh with at least ¼ inch of cement. To make the cement stick well and be easy to work with, it has to have just the right amount of water, not too much and not too little. The only way to learn is by doing it. Mix the sand and cement dry and add the water slowly. If you don't have

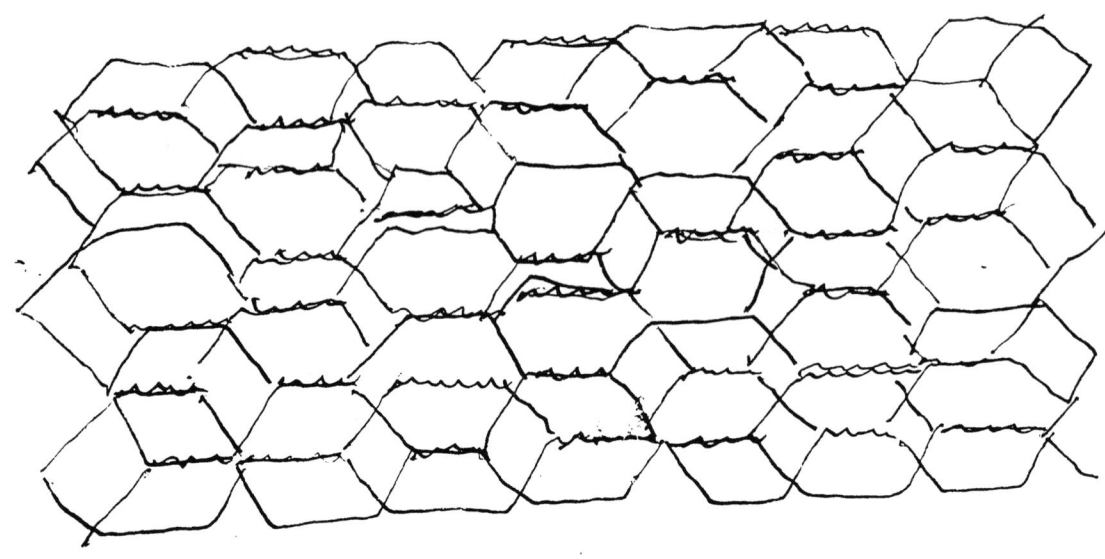

When tying wire mesh on always try to overlap it in such a way as to make the holes smaller. This will make plastering much easier.

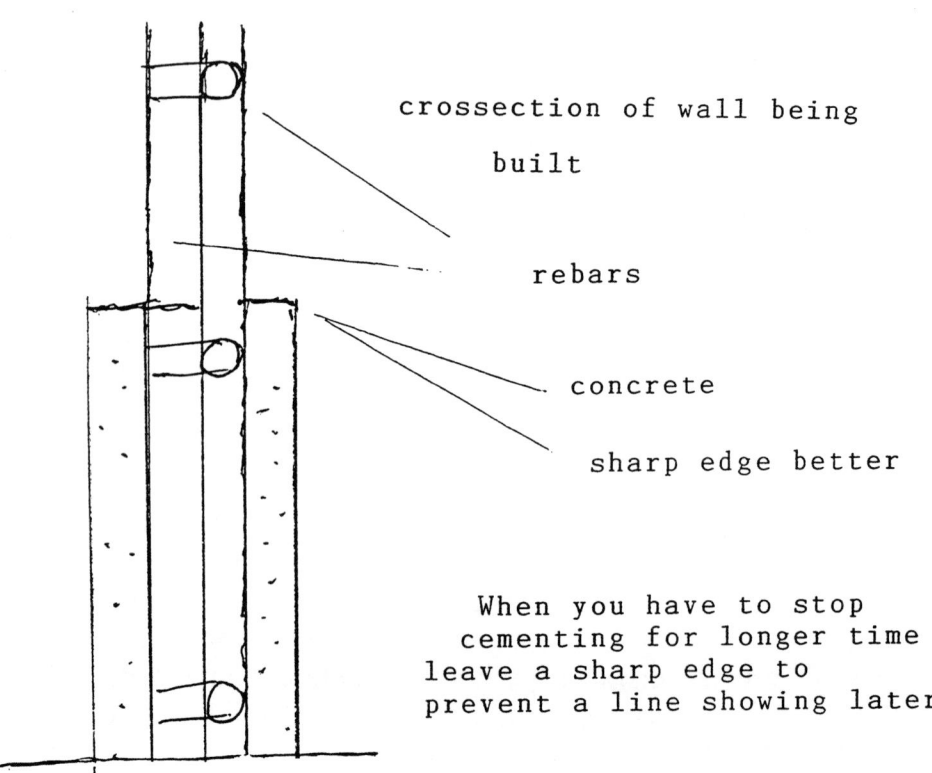

crossection of wall being built

rebars

concrete

sharp edge better

When you have to stop cementing for longer time leave a sharp edge to prevent a line showing later

enough workers to finish cementing the whole dome in one day and you have to stop and start the next day, one way to prevent unsightly lines where the old cement meets the new cement is to make the edge of the last layer sharp enough, this will also help you to see the thickness of the cement over all bars and wire mesh, which should be the same all over the dome. Approximately quarter inch.

Try to avoid rounding the edge as that would make the next layer overlap the lower one and it might show that line. Another way of doing it is like this, you could rough coat the whole dome with one layer and then make one final smooth finish coat in one plastering job, if you do it that way, make sure to scratch the first coat when it is wet with many vertical and horizontal lines like an XXXX. Make the lines just an inch or two apart so it is quite rough and the second final coating will really stick. You have to keep wetting the rough dry coat well, too.

I like the first way better though because unless you have many good plasterers on hand, you will never get a big dome done in one continuous run, so you will have to stop anyway.

For cement to reach its ultimate strength, it has to be wet cured. So keep it wet continuously. The first three days are most critical, but keep it completely wet for at least seven days.

Actually, cement keeps curing and hardening indefinitely but after 30 days it is almost to the maximum strength, like 90% or so, so it helps to keep it moist for that time. This is especially important in a thin wall structure like a dome.

You could also quick cure it with hot steam, if you cover the whole structure with a plastic sheet, make it airtight, and get a hot steam machine working inside, it would be almost completely cured in three days.

It is something to think about and seriously consider if you happen to be working in winter when it is freezing outside. If you don't keep the whole dome wet, it may crack some place so make sure to always have plenty of water in reserve and keep wetting it.

By the way, if you want to sound like a pro when discussing the job with other trades people, the mix of sand, cement and water is known as a mortar mix , only when the stuff hardens it becomes concrete. Don't confuse the two, if you do something, do it right.

The easiest way to insulate your dome is to have someone to spray on the expanding foam insulation. I do not want to mention any trade names, as I don't wish to advertise something that may not work to your satisfaction. So use your own judgment when using a product, whatever it is, always ask around for satisfied customers.

The cheapest way is to do it yourself, using 2 x 8 or 4 x 8 sheets of rigid insulation. Some are easy to bend, others not, so use the thinnest ones and overlap them. Put as many layers as you need depending on where you live.

Residents of Alaska or Northwest Territories will obviously need thicker insulated houses than those living farther south in comfortably warmer climates. However, if you live too far south, in a tropical climate, you will need almost as much insulation just to keep the house cool. Keep in mind though that the concrete dome itself and if you have a concrete floor, it will store and hold the same temperature for a much longer time than the regular wooden house, so you will not need as much insulation as a stick 2 x 4 house.

You can start putting on the insulation as soon as the concrete sets. By covering the outside of the dome's concrete completely with waterproof insulation, you will keep the moisture in, so it will still keep curing well. Keep spraying water on the interior side for at lest seven days though, to be sure the concrete cures well without cracking.

Outside Dome Cover. Forget about shingles or plastic siding, we will use something better, you guessed right, concrete. Once the dome is all plastered over and you have all the insulation on, you will have to cover it so it is protected from the weather by putting on a layer of cement.

While the cement sticks actually pretty well to styrofoam it still should be reinforced with wire mesh. Best wire for this job is heavy fencing mesh. You know, the one that is used for fences around houses, parks, schools and almost everywhere else.

Put on just one layer and make sure it is evenly and smoothly placed all around. Wherever you have to put it on a vertical wall you must leave plenty of short tying wires running from rebars on the interior wall through the insulation to the outside, otherwise your outside mesh and cement will have nothing to hold it in place. Plan ahead and put enough of these tying wires wherever required before you do any cementing.

You don't have to put any of these tying wires on a dome, as the weight of the cement will keep the mesh in place. When cementing this final coat, try to do a nice even layer, making sure to cover all the mesh well. At least half an inch should be plenty to protect the insulation from anything.

Make this final coat very smooth all over as that is what you will be looking at for a very long time.

If you can't make it smooth for whatever reason, you could also make some ornamental plaster finish or even hand prints or paint some nice mural on it. You can also set a bunch of little stones or pebbles into the wet cement, or lay on ceramic tiles of different shapes and sizes and colors for a true artistic expression, or you could just plant some vines close to the dome and let it grow all over it ...

As you can see, the possibilities are endless and all domes need not look alike.

Vertical wall

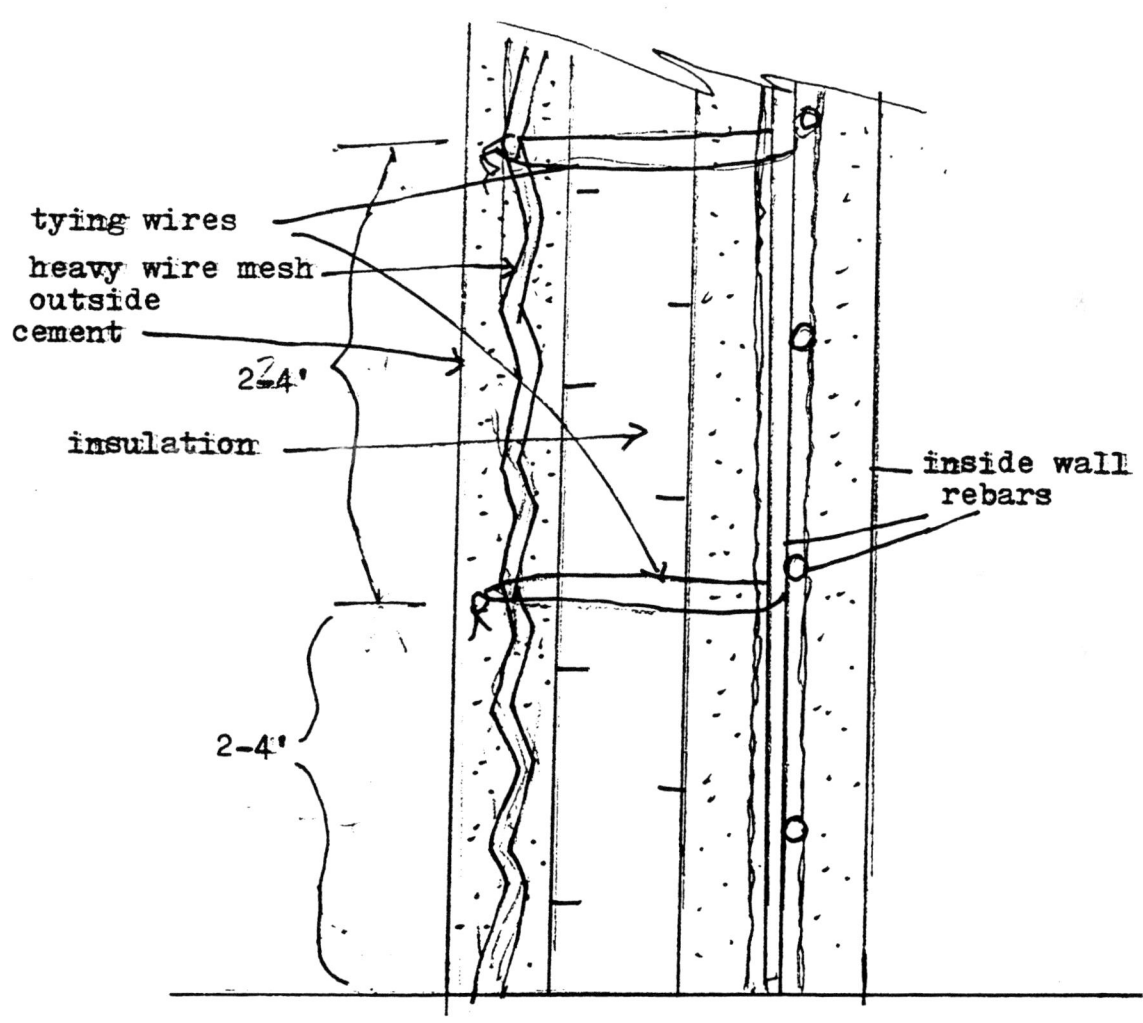

Building a Vertical Flat Wall from Cement

Sometime it is advantageous to build a wall flat (vertical) instead of round. Such as where you have to place doors or for interior partitions. Doing these dividing walls is fairly simple. You have to plan these when laying the concrete slab, which could be after your dome walls are up, for dome walls you only need footings to build on. If you have a wooden floor, you will have to do wooden inside walls also.

So if you want to do inside walls from cement, too, you must tie their rebars to the dome before you plaster it. Something to decide on very early when drawing your plan.

One good thing, at least you don't have to insulate these walls, unless you want to make them soundproof. That is one disadvantage, thin concrete without insulation is not a very good noise stopper, the same as plywood or drywall.

Fortunately the outside wall must have insulation and rigid foam is one of the best insulators, so don't worry you won't hear much traffic noise unless you live near an airport runway.

If you need to attach insulation to a flat wall such as garage end walls, use the rigid foam again, just tie short pieces of tying wire from rebars past the insulation and leave it long enough to tie to the wire mesh, which needs to be placed over the insulation on the outside and plastered over to protect it from the weather.

Tie these wires every two feet or so vertically and horizontally. If the insulation gets in the way, just slice the wire into it, watch it so you don't slice your fingers, and tie on the wire mesh.

The simplest way to attach tying wire is to double it, so it looks like a very long U, then just slip it over the bar, pull both ends even length through the wire mesh and twist these together, don't pull too fast or you will cut your hands. Easy does it.

Second Floor

The second floor or loft and any other partitions for all the rooms can be made just like any regular house, yes, from wooden 2 x 4s or metal studs and drywall, or bricks or ferrocement or even logs, if you want a break from all that cement work.

If you need to attach the wooden studs to a concrete wall, use a nail gun that shoots nails through the wood into the wall. To make a straight wooden 2 x 4 stud conform to a curved wall, you can slice about three quarter way through it every couple feet or so and it will bend real easy. If you are planning to make a bathroom on the second floor, make it directly above the main floor's bathroom, it will make installing the water lines and vents so much easier. Try never to place bath above kitchen, I'll explain why later.

Stairs

Round stairways are most efficient as they use the least space. If you want to make a long straight, curving or some other fancy design, it can also be done. Just your imagination and financial situation set the limit. Only make sure that you have enough space for it. Stairs are best placed somewhere close to the middle of the dome to keep the design simple. The curvature of the dome's wall requires the stairs to be curving away from the wall towards the middle of the dome as you go up. It can be done with a little planning, but it is a bit more complicated than a straight stairway.

Avoid thermal breaks. No, this has nothing to do with taking a break to drink from your thermos bottle, it means that you have to make sure to separate the dome's interior wall from the outside concrete wall by insulation. Everywhere. Do not overlook any corners or foundation. If you forget to place insulation even on the smallest patch of wall, you will have a thermal break, meaning the cold air will penetrate to the interior wall and when inside warm air hits the cool surface, the moisture will condense on it and it will be dripping wet as long as it is cooler outside than inside. That is why we also have to use double pane windows everywhere. It would not be really a big disaster because concrete wall doesn't get damaged by water drops, it would be more of an inconvenience. So make one continuous insulating cover all the way around the dome.

Doors and Windows

When it comes to windows and doors, there are several ways to go. Windows can be made round, square or any other shape you can think of. You could even use the same windows they use in customized vans, diamond, heart shape or what have you, as long as they are double pane for insulation and the frame is wide enough to fit the wall. You could also use skylights, some of them open, some don't. I would not put the ones that open on top of the dome, you would have to remember to close them every time you leave the house so in case of a sudden rainstorm you wouldn't get some water damage inside on your furniture. Doors will obviously need a flat section of the wall unless you want to create some custom design curving the same way as the wall.

   I like to make these dormers like extensions, even though they will take a bit more work, I think they look better than those recessed ones (see picture) and are more practical as they give you a few extra feet of space. But it is your design, so choose and use whichever you like best.

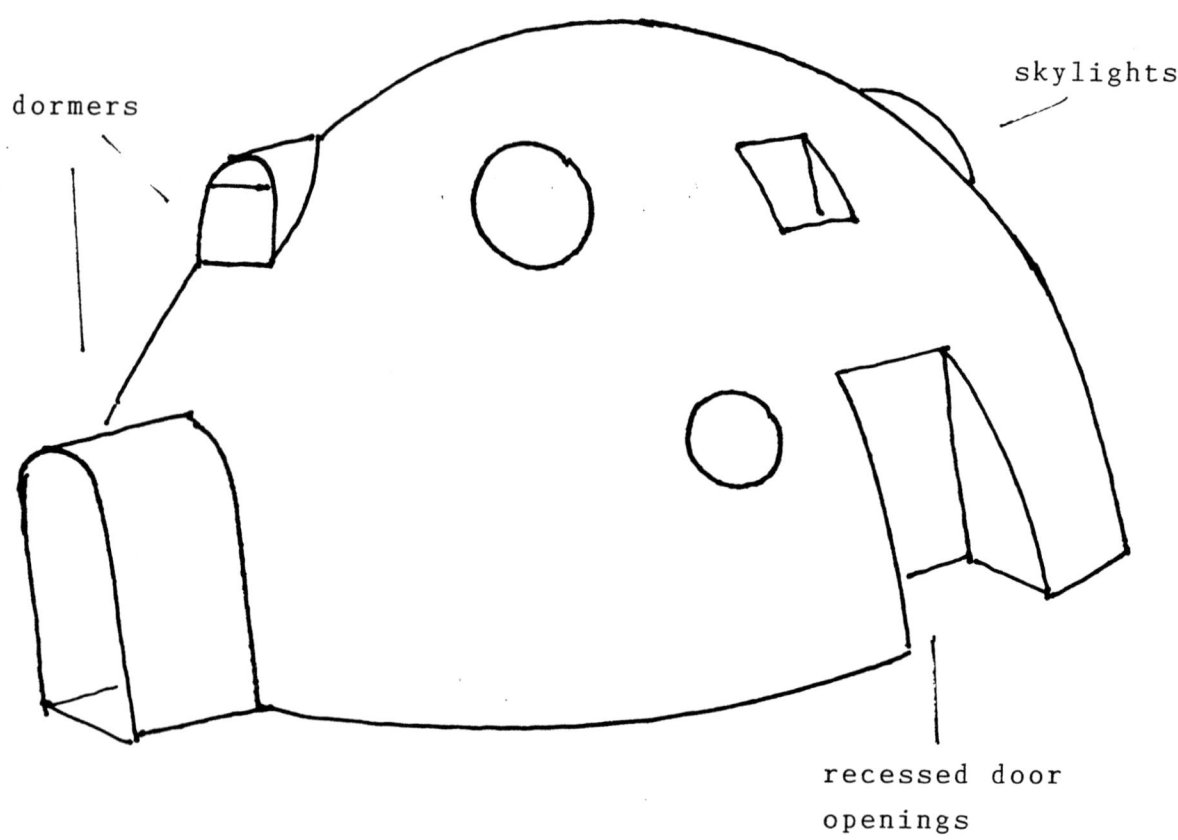

So why should you never place a bathroom above a kitchen? Some people consider it bad karma, bad luck or plain bad vibes. If you would ever want to sell your house, an oriental person would never buy it, if there was a bath above the kitchen. Why? It is called harmony with nature (or something to that effect). In Chinese Feng shui. Everything in and around the house has to be built and arranged in certain, harmonious ways so that people living there will be happy and have good luck. It makes sense in many ways. For example, you should never have the front door facing directly onto the street, too much negative energy rushing in and upsetting people, if noise is negative, that is certainly true.

Also, you should not place water such as sink opposite or next to the fire such as an oven. Water and fire do not mix, so obviously they should be far apart.

I think it is more of a common sense from the olden days when people used to have fire next to water if you accidentally spilled the water and put out your fire, it would be quite inconvenient, or maybe even dangerous, int he winter you could freeze if you had no matches and dry wood to start a new fire. Now it seems more like superstition. But if you had a bathroom over the kitchen and your toilet accidentally overflowed just as you were cooking some nice big fancy feast for someone very important, you can imagine how bad the vibes or luck or embarrassment could get. Makes sense, doesn't it?

Another thing, to be safe and comfortable when working in office let's say when sitting at a desk, you should never have a door at your back, I guess so no one could sneak up on you, too bad that Wild Bill Hickock did not know about this Feng shui business. Bad luck or what? Also, you should never place your bed and sleep under an open window, but I just can't figure out why, can you?

Now I don't know all the details and regulations or restrictions of this science, if I can call it that, there are many other books out there that will give you more information if you are interested in this.

I think lots of it is common sense and some is a bit of superstition. But make what you want of it. Everybody wants a house that is comfy and happy to live in. And when it comes to luck, I just don't believe in it. I believe that your life is what you make it. Like they say in the Army: A good soldier doesn't need luck. If you believe in yourself, you can do and accomplish anything. I met a pilot once who had a sign on the dash that said: God is in charge, I am just a copilot. With an attitude like that I wouldn't trust him to drive or fly me anywhere. When you drive you better be 100% in control or as we flyers say: Pilot in command, you are responsible for your actions (or at least think you are), do it any other way and you won't make it, leave the negative baggage (thoughts) elsewhere and pay attention to what you are doing. But enough of the philosophy crock, back to building domes.

Short History of Ferrocement

Ferrocement goes back all the way to dinosaurs. What, am I kidding? Yes, I am, but I am sure you have seen them too, along the road. I don't remember exactly where, either California or somewhere close there, someone has built a bunch of dinosaurs and some other weird looking prehistoric creatures from ferrocement. Why? I think it is a park or they found remains of them or something. It is just a good example of what can be built from ferroconcrete.

Actually ferroconcrete goes back at least a hundred years or so when someone first built a small boat out of it, guess what, it still floats.

Before the Second World War, there were quite a few big ships and barges built using ferrocement. Today people use it mostly to build swimming pools, retaining walls, very few boats, yes, even canoes and the occasional dome home. Usually underground, I guess most people are not too hot for concrete. I don't understand why, with all the advantages concrete has. And if you don't like its color, you can always paint it a different one.

Now you might wonder how do I know it will be safe in an earthquake. During some construction work, we had workers trying to break reinforced concrete with jackhammers and it took incredibly long and strong pounding and hammering before they smashed the concrete, however, the reinforcing rebars still held most of it together until they cut them with oxy-acetylene torches.

As for concrete boats, I have heard the same thing, if the boat accidentally hit some reef or rock so hard that it cracked the concrete, the wire mesh still held it all together without any catastrophic consequences. Even though some water kept coming in, bilge pumps were fortunately working well and the boat made it safely home to port, without sinking.

So it makes sense, when earthquake shakes a regular house, it will break it apart, usually at the corners, they seem to be a weak spot. But in a concrete dome reinforced with many rebars and wire mesh this won't happen, you would probably get some cracks no doubt, but it will not fall apart like some stick house. So you don't have to worry about the roof falling on your head and crushing you.

The roundness of the dome makes it self-supporting without any corners it has no weak spots as long as you don't make some huge windows or doors. If you do, make sure to increase the reinforcement bars around these big openings.

Ferrocement is best.

A Few Words on Hiring Workers and Helpers

If you cannot or don't want to build the house yourself, you may have to hire subcontractors, carpenters, electricians, plumbers, concrete workers and labourers to do some or all of the work for you.

If you go this way, always make sure to get references from satisfied customers. And always make a written contract for any big job, should there ever be some screw-up and you needed to go to court to get a judge to settle the disagreement, your contract is your guarantee that will prove who is right and who is not.

In court one person's word is just as good as another's (usually). Unless you know the person doing the work for you very well, write down that contract, it doesn't have to be anything complicated just put down exactly what is to be done, for how much money in what time schedule, and most important, to your or the building inspector's satisfaction.

If the worker is good, he will usually agree without hesitation, assuming the demands are reasonable. If someone is crooked or clumsy, you will hear a whole lot of excuses, b.s. or demand for a down payment such as permits etc. Do not fall for any nonsense like that.

Never give any worker down payments or a deposit before the work is started. If the subcontractor doesn't have enough money to work with, he or she is probably not very reliable. And if you give them money before the job is started you may never see it finished.

Do not never ever get influenced by anyone's sweet talk and promises. Be polite but firm, tell it like it is: Get the job done first and to my satisfaction then you will get paid. Case closed. And make them sign the contract, always.

Follow Your Instinct, Your Gut Feeling

You can usually tell by a person's talk, behavior and/or body language whether he or she is a professional tradesman or a pro hustler. The first one you want to do the work for you, the other you politely show the door. By the way, the subcontractor buys the permit, he will bill you later when the work is done but never before.

Another important advice I would like to share, when buying land always use a reputable lawyer, no matter how easy or simple it looks, it can get pretty complicated. Trying to save a few hundred bucks might end up costing you much more if something is not what you thought it was. Lawyers know, or at least should know, what has to be done to the last detail. If he screws up, you will sue him, if you screw up, you are what? Get a lawyer.

Estimating the Cost

The cost of a concrete dome will vary only slightly from one area to another, depending on prices of your local building materials. Cost of interior finishing, fixtures, appliances and furnishings will obviously be quite different from house to house as some people have and like to spend more money than others.

Fortunately cement and sand is still reasonably low priced almost everywhere. So that concrete dome can be built for less money than any other comparable house of same size.

Everyone's interior decorating ideas and spending habits are different, so I won't even try to guess what you want the inside of your dome house to look like.

I can only give you a rough estimate on cost of the dome shell. Without windows and doors, as these can vary in price form $20 for second hand ones to $200 or more for brand new ones.

To estimate cost of dome walls, you have to figure out a square footage of toal surface of dome.

Not a very easy thing to do, but you will come pretty close if you measure the circumference (meaning all the way around) times the height to the top. Along the wall, straight up. However, as the dome narrows on top, it will need less materials, such as insulation boards. In case you miscalculate slightly without knowing it, just to be safe, you don't buy too much material, buy only half or even less and when you use it up, get some more.

Starting with the floor. A concrete slab which is usually four inches thick, plus a little deeper for wall footings. For example, a 26 foot diameter dome will use about 10 cubic yards of cement. One cubic yard is 3 x 3 x 3 feet. It costs $85 to $95 delivered, that is $950.

If you mix it yourself, the price will be around $300. The choice is yours, do you have more time or money?

If you need stronger reinforcement bars, such as for a garage where you will drive heavy vehicles, add the cost to it. If you need labourers to help you build it, add some more. Rigid insulation, one 4 x 8 sheet, one inch thick, costs $8. Placing these along the outside foundation, you will need ten sheets. 10 x 8 = $80.

A 26 foot diameter dome will use approximately 140 rebars, 20 feet long. The cost is $5 for a ½ inch bar. Thicker bars obviously cost more. 140 x $5 = $700.

The surface of the dome is about 1900 square feet. If you make a two inch wall, you will need 12 cubic yards of cement.

You should mix this yourself on site as you will only need small batches at a time, unless you have a big crew of plasterers. One cubic yard of concrete will take 3 bags of cement plus sand and water. One bag of cement costs $8. That is $24 plus, maybe $25 at the most, for sand, lets say $30 a yard. A 26 foot dome uses about 12 cubic yards of mortar, which is $360.

Wire mesh. Chicken wire mesh costs $12 per 75 square feet roll. We will need 25 rolls of mesh for one layer. We have to use at least two to four layers, you might try to get away with three or even two layers if your rebars were really close together but the plastering would be quite a bit more difficult, especially on top of the dome when the wall starts curving over and becomes ceiling, so the more mesh you use, the easier it will be to plaster. So four layers of mesh will be better in every way, cost $1,200.

Insulation. Using one inch of 4 x 8 sheets of rigid insulation, the dome will need 60 sheets at a cost of $480. Two inches insulation will be $960. If you need thicker insulation, it will obviously cost even more. I doubt it will be really necessary.

Heavy wire mesh (chain link) for outside wall costs $55 per 200 feet square roll. As we have to put just one layer, we will need just six rolls of mesh, at a cost of $339.

So, when we add up all the materials, slab foundation $958; rebars $700; cement $360, wire mesh $1,200, insulation 2" thick $960; chain link mesh $550; sand $200 ... altogether $5,000.

To this you will have to add the price of plumbing, electrical wires, doors, windows and maybe more rebars if you need them thicker and/or closer together. I have estimated 12 inches apart, for a 26 foot dome it should be plenty. However, to be sure, get your design approved by a reputable engineer.

If you have more and big windows, you will need more reinforcing rebars and closer together, I am almost sure.

Estimating cost of labour is the tricky part because it varies so much from one area to the next. And from one worker to another. Get a Union help and you will be in the poor house very quickly. Do it yourself and you can afford a mansion. Or at least a nice comfortable low cost, energy efficient, almost indestructible shelter.

For around $10,000 for a 26 foot dome, which gives you about 600 square feet of floor space, that is $17 per square foot and a pretty good deal anyway you look at it.

Now, a 26 foot dome may be OK for a garage or a single person home. With a loft or second floor, you could even have two people living quite comfortably, but for a small family you will need at least a 40 foot diameter dome.

So here is the estimate for a 40 footer. Building a four inch concrete slab, which is about 1,600 square feet big will use 20 cubic yards of concrete, at $95 delivered, totals $1,900, or about $800 to do it yourself.

Insulation for slab, 28 sheets of 4 x 8 boards at $8 each x 28 = $224. If your frost line is deeper than four feet, you will need more sheets of rigid insulation.

For a dome this size, I would use at least 5/8 or even bigger rebars. These are $9 for a 20 feet long bar, we will need 224 x $9, which totals $2,000.

Concrete for the dome itself will take 18 cubic yards, if you mix it yourself, it is $800.

To cover a 40 foot diameter dome with four layers of wire mesh, you will use 166 rolls at $12 a piece, totaling $1,992.

Heavy wire mesh costs $55 per 200 feet square roll. We will need only one layer which will take 10 rolls, cost $550.

Using rigid foam insulation, 2" thick, we will need 196 four x eight sheets, these cost $1,568.

Adding up all the numbers, the basic dome shell will cost $9,000. To this, you will have to add the cost of windows, doors, electrical wiring and lights, plumbing and water and sewer lines, and most likely labour, as building this size of a house is quite a bit of work to do alone.

Also you will want to build interior partitions and a second floor as there is plenty of room for at least one bedroom and bath upstairs.

All this will probably cost anywhere form $15,000 to $20,000 or maybe even more. It all depends on how expensive your tastes are and how cheaply you can buy your materials and labour.

I am sure there are people who will easily spend hundred thousand dollars or more but for those on a limited budget, the concrete dome can be quite affordable. And the best thing about it is, it should not need any repairs or big maintenance for a very long time. And should your dome ever be in the middle of a tornado or a hurricane, I am sure it will still be standing after the storm is gone. How many houses can you say that about today? Only the concrete ones.

The tools needed to build a ferrocement dome. It all depends on how much of the work you will be doing yourself. For you financially independent, all the tools you may need is a pen to write checks. For those less wealthy but handy enough to swing a hammer without knocking yourself out (it has happened believe it or not), you will first of all need two strong hands (or four or more) and some knowledge of plastering and cement work would help too. For foundation and interior work, you will need the usual assortment of carpentry tools, hammer, saw, level, angle, crowbar, tape measure, pliers. Use power tools as much as possible for easier work, if you don't have them, you will have to use your muscle power, either way works and can be done.

For cement work you will need shovel, cement mixer, wheelbarrow, trowels pointed and square, metal and wooden float. Also, scaffold, stepladder, buckets for water and for cement - use the square ones with low sides for easy scooping of mortar.

For metal work you will need wire cutters, rebar cutter, either a hacksaw or a grinder with cutting wheel, or oxy-acetylene torch, pliers.

You may also need protective gloves, for handling metal rebars and rubber boots for walking in cement, such as when making concrete floor. If you get cement on leather shoes, you can pretty soon throw them away as wet cement will quickly destroy leather. Also have a good supply of stick-on bandages. You will certainly need them.

Other than that, that's pretty much it. So have fun and enjoy building your dome.

The End.

CROSSECTION OF FERROCONCRETE DOME WALL
(not to scale)

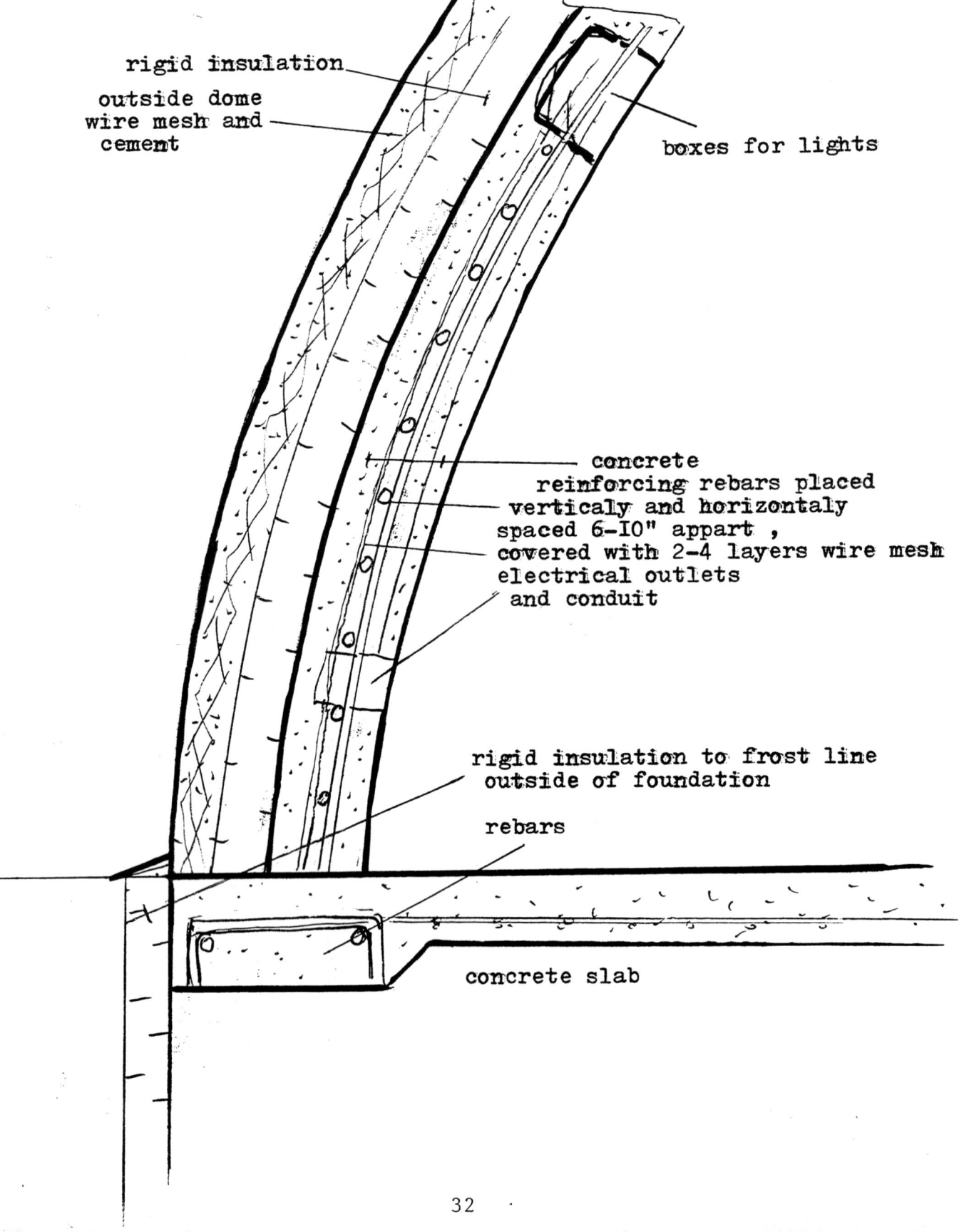

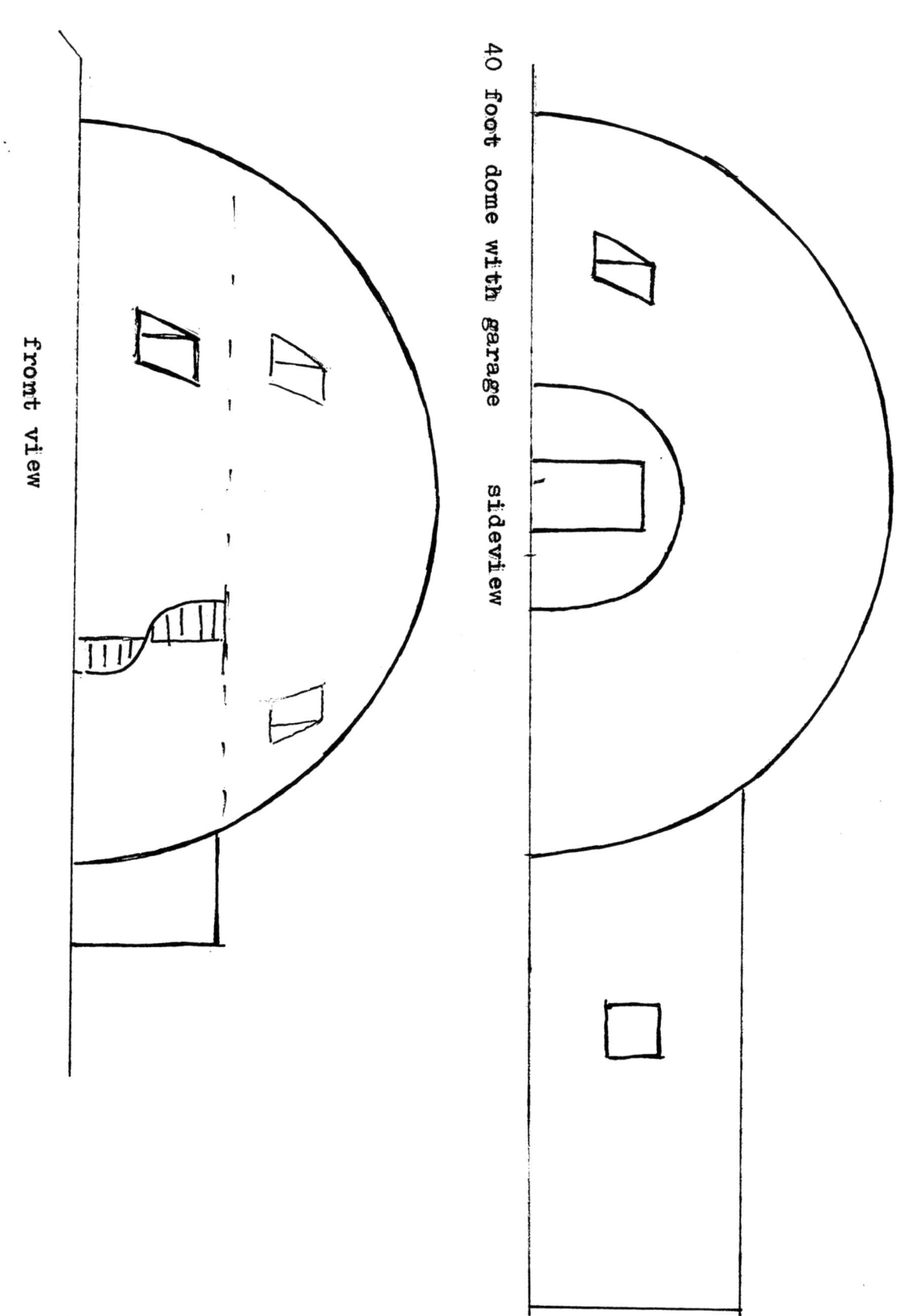

40 foot dome with garage   sideview

front view

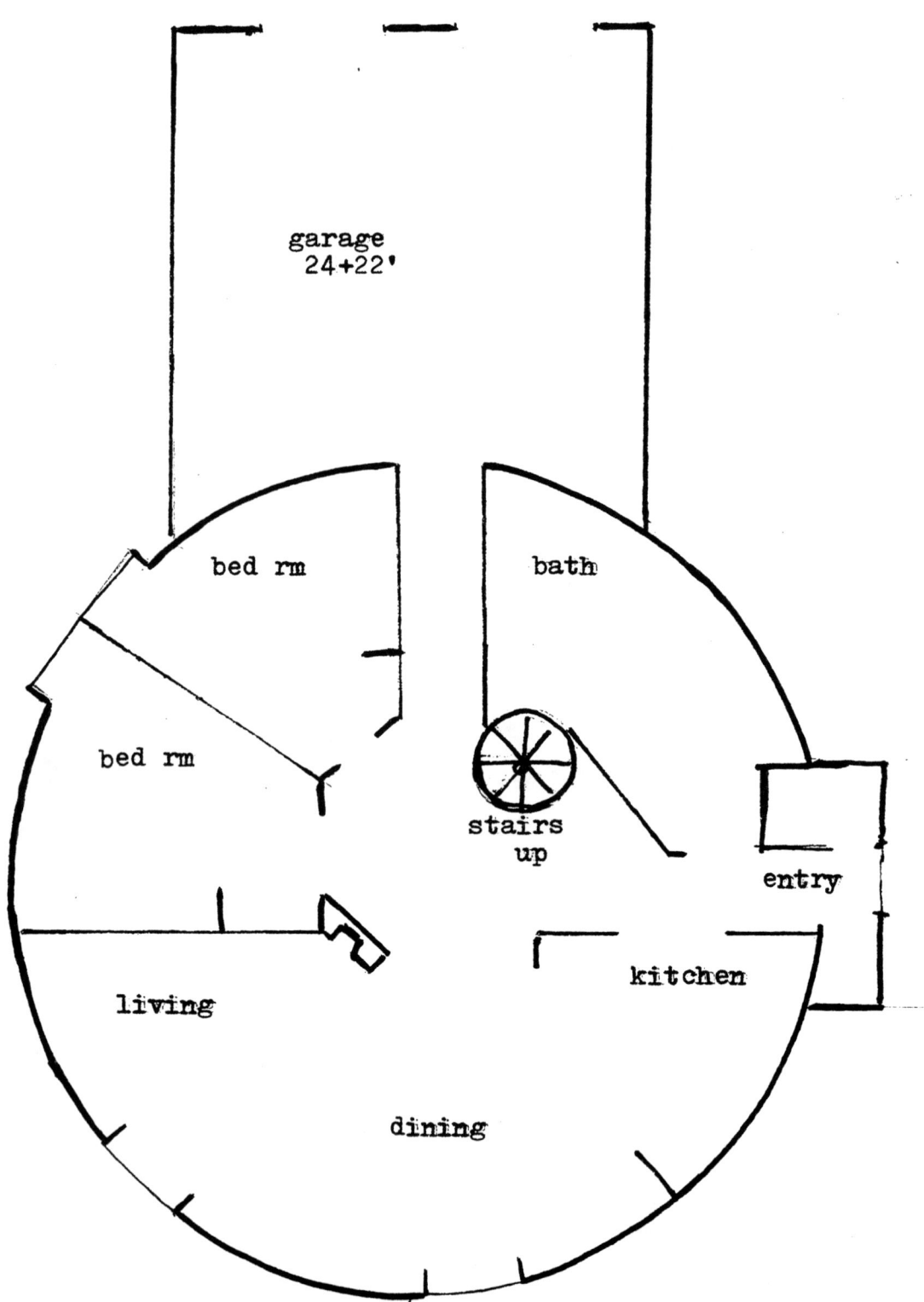

main floor 1500 square feet

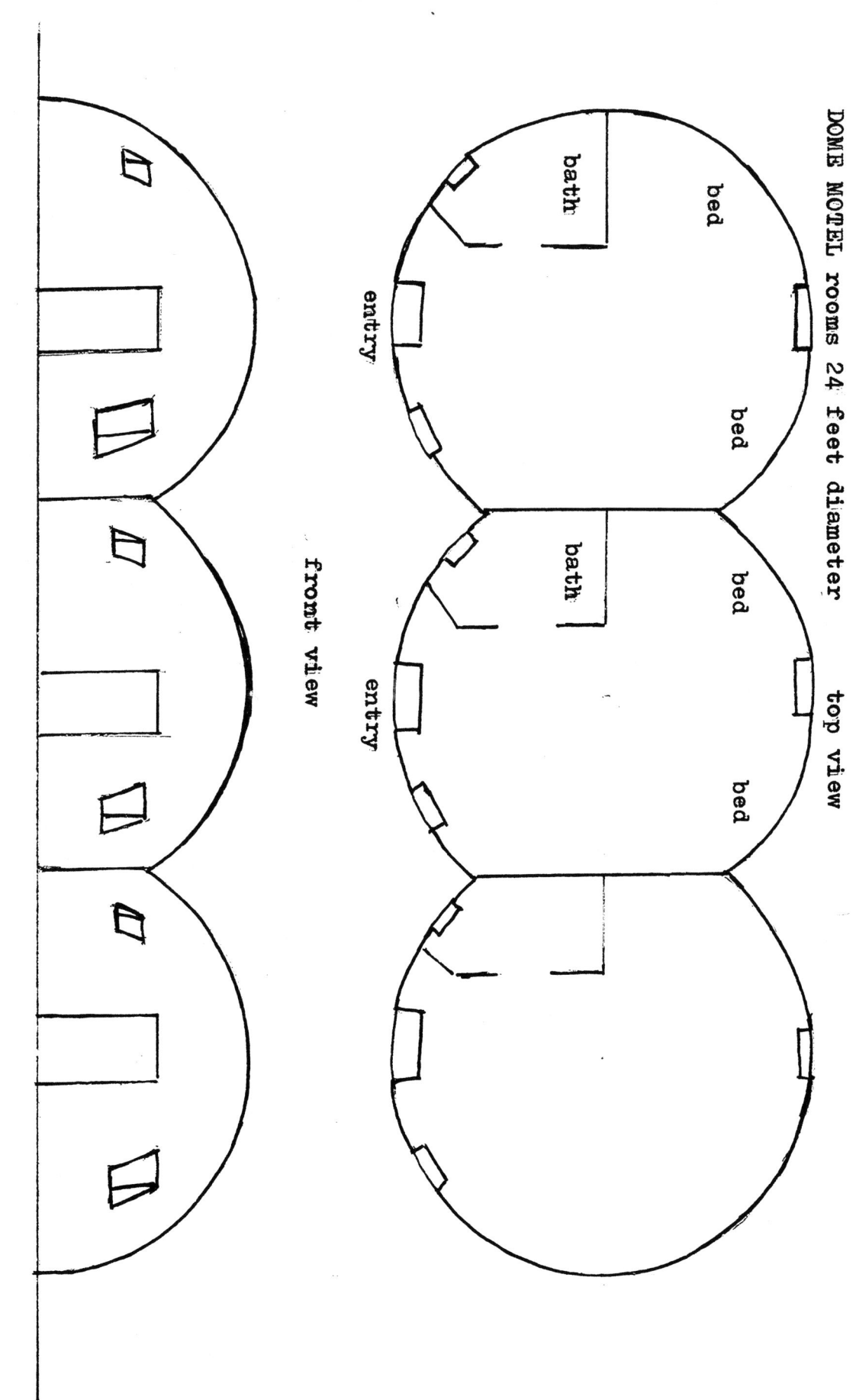

2 CAR GARAGE WITH LOFT

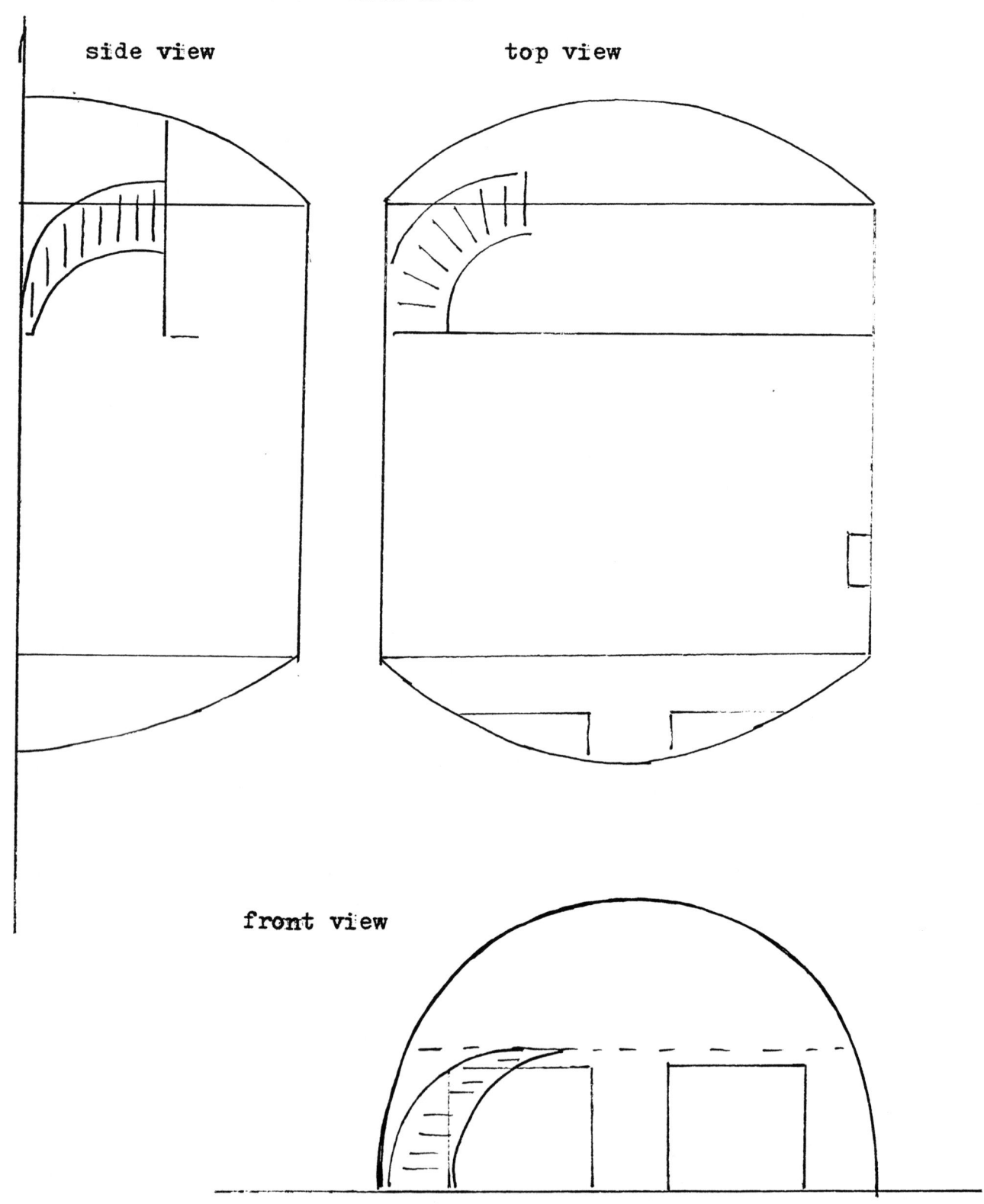

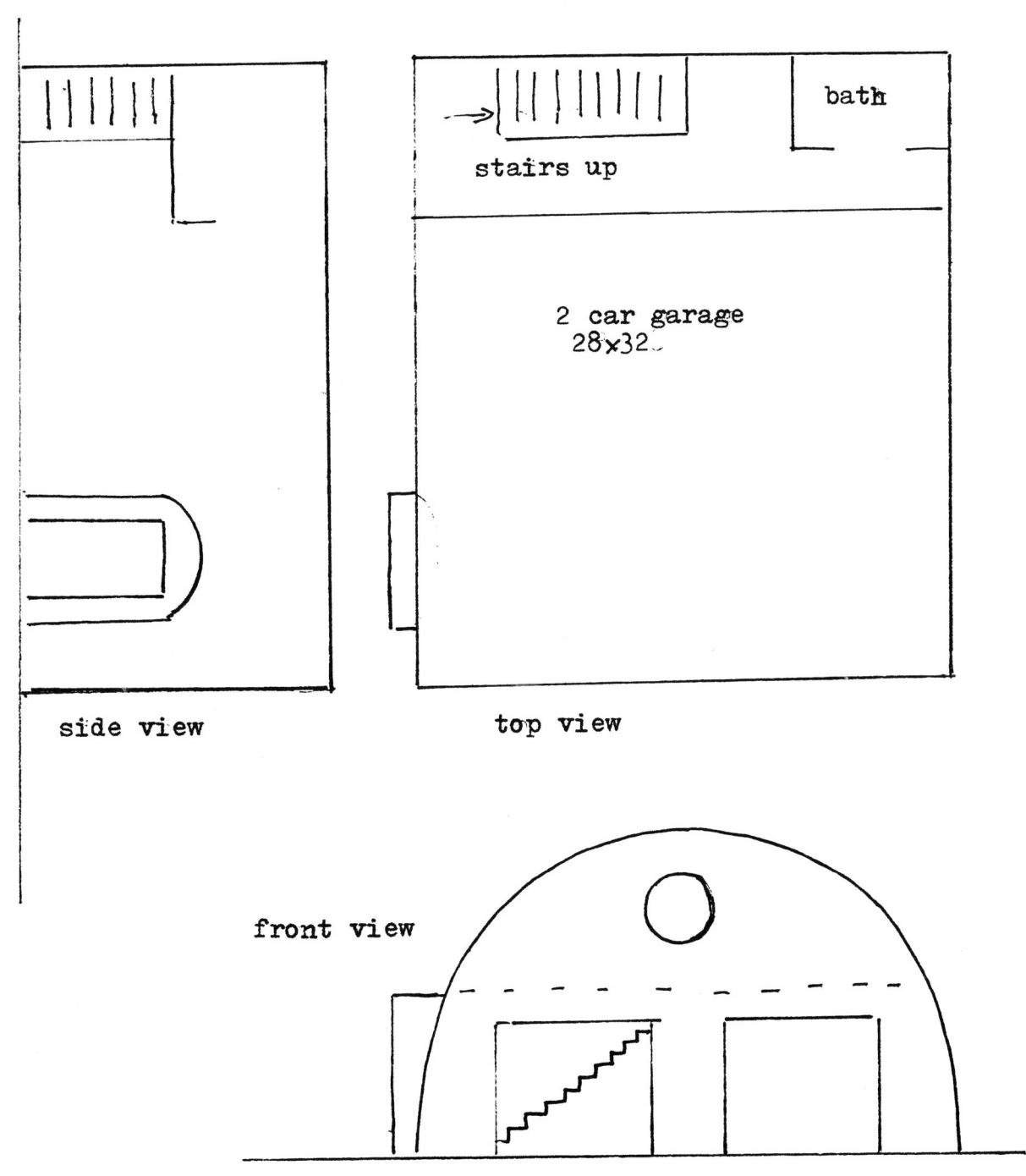

2 CAR GARAGE WITH LOFT ABOVE,
MAIN FLOOR GARAGE, WORK AREA, BATHROOM

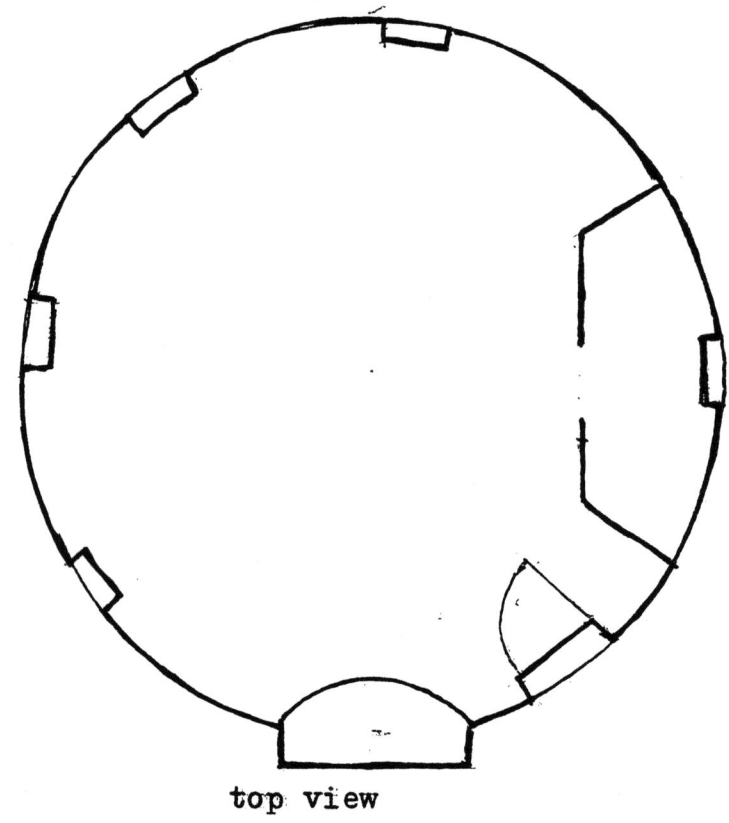

top view

front view

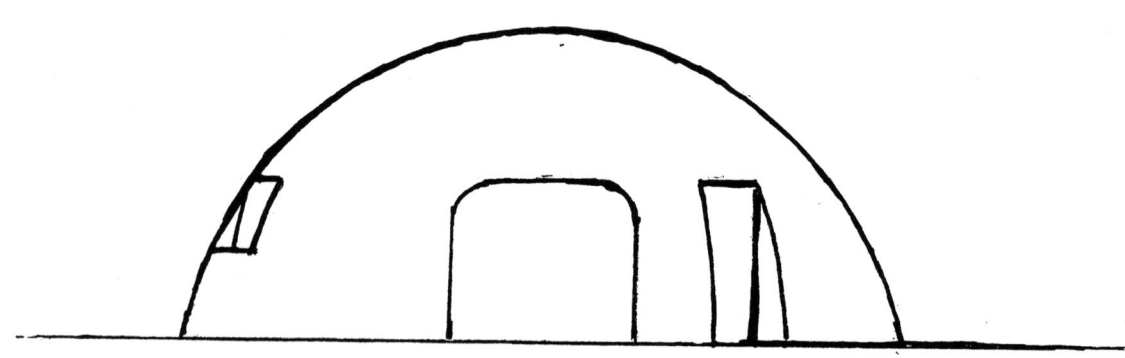

1 car garage,workshop,with bath .30' diameter.